FLORA OF TROPICAL EAST AFRICA

———

PHYTOLACCACEAE

R. M. Polhill

Woody or herbaceous, erect or scrambling. Leaves alternate, simple, entire, pinnately nerved, with crystals generally easily visible in the younger leaves. Stipules generally lacking. Inflorescences racemose, spicate or paniculate, terminal, lateral or axillary; bracts and paired bracteoles small. Flowers small, ☿ or ♂ and ♀ (then rudiments at least of aborted organs still present), usually regular. Sepals 4–5 (rarely more), free or partially united, imbricate, usually persistent, sometimes brightly coloured. Petals generally absent (present only in *Stegnosperma*). Stamens (3–)4–many in 1–several whorls, when definite alternate with the sepals, usually inserted on a ± fleshy disk; filaments free or united at the base; anthers 2-thecous, dehiscing lengthwise. Ovary superior or rarely semi-inferior, of 1–many free or united 1-ovulate carpels; styles as many as the carpels, free or connate at the base, short, sometimes lacking; stigmas linear or capitate; ovule basal, shortly stalked, campylotropous. Fruit of 1–numerous free or united carpels, fleshy or dry, sometimes winged. Seed subglobose, discoidal or reniform, with a membranous or brittle testa, occasionally arillate; embryo annular, surrounding the endosperm.

A predominantly American tropical and subtropical family of the order *Centrospermae*, comprising 16 genera and some 100 species following Heimerl in E. & P. Pf., ed. 2, 16C: 135–164 (1934). The family has been further subdivided by Hutchinson (Fam. Fl. Pl., ed. 2 (1959)) and Airy Shaw (Willis, Dict. Fl. Pl. & Ferns, ed. 7 (1966)), *Hilleria* and *Rivina* being placed by Hutchinson in the *Petiveriaceae*.

Ovary of 5–16 carpels; fruit a lobed berry . . . . 1. **Phytolacca**
Ovary of 1 carpel:
  Calyx with 3 anterior sepals united to ± the middle, the
    posterior one free; fruit dry . . . . . 2. **Hilleria**
  Calyx of 4 free subequal sepals; fruit juicy . . . 3. **Rivina**

1. PHYTOLACCA

L., Sp. Pl.: 441 (1753) & Gen. Pl., ed. 5: 200 (1754); H. Walt. in E.P. IV. 83: 36 (1909)

Trees, shrubs or herbs, erect or scrambling. Flowers in terminal and lateral racemes, spikes or panicles, ☿, sometimes dimorphic, or ♂ and ♀, regular. Sepals (4–)5(–6), free, usually subequal, often becoming reflexed. Stamens 5–33 in 1–2 series, sometimes with the filaments shortly united at the base; anthers dorsifixed; stamens of ♀ flowers reduced to varying degrees. Ovary ovoid to subglobose, with 5–16 free or united carpels. Fruit a lobed berry, rounded or depressed-globose, thin-skinned. Seed lenticular or subreniform, with a black brittle testa.

About 35 species in tropical and subtropical regions, mostly in America.

P. dioica L., a South American species, has been grown as an ornamental tree at various places in East Africa, e.g. Uganda, Entebbe Botanic Gardens, *Chandler* 1143 !,

Kenya, Marsabit, *T. Adamson* in *E.A.H.* 11304!, Nairobi Arboretum, *G. R. Williams* 329!, Tanganyika, Morogoro, *Wigg* 976! Dale, Introd. Trees Ug.: 55 (1953) is rather disparaging of it. Apart from its habit, it may be further distinguished from *P. dodecandra* by the 20–30 stamens, the broader glabrous sepals, the more marked differentiation of the flowers (the staminodes being quite small or lacking in the ♀ flowers) and the fruits with less divergent carpels.

Flowers dimorphic, the carpels of long-staminate flowers
    rarely developing; carpels (4–)5(–8), united only at
    the base; fruits bluntly lobed with the style-bases
    at the points, ripening orange or red   .   .   .   1. *P. dodecandra*
Flowers all alike, the stamens not exceeding the sepals;
    carpels (7–)8(–10), united; fruits depressed globose,
    only slightly lobed, with the style-bases around a
    median depression, ripening black or purplish   .   2. *P. octandra*

1. **P. dodecandra** *L'Hérit.*, Stirp. Nov.: 143, t. 69 (1791); Bak. & Wright in F.T.A. 6(1): 97 (1909); H. Walt. in E.P. IV. 83: 42, fig. 15 (1909); V.E. 3 (1): 137, fig. 87 (1915); T.S.K.: 14 (1936); F.D.O.-A. 2: 253 (1938); F.P.N.A. 1: 144 (1948); T.T.C.L.: 449 (1949); Balle in F.C.B. 2: 94 (1951); F.W.T.A., ed. 2, 1: 143 (1954); F.P.U.: 106, fig. 55 (1962); F.F.N.R.: 46 (1962); Verdcourt & Trump, Common Pois. Pl. E. Afr.: 29, fig. 2 (1969). Type: plant cultivated in Paris from seeds collected by Bruce in Ethiopia (P, holo.)

A semi-succulent straggling or scrambling shrub up to 4 m., sometimes climbing up to 6–9(–18) m., glabrous or less commonly crisped pubescent on the vegetative parts. Leaf-blades ovate to elliptic, 6–15 cm. long, 3–8·5(–10) cm. wide, narrowed to a prominently and ultimately recurved mucronate tip, cuneate, rounded or cordate to the shortly decurrent unequal-sided base, glossy, with translucent margins (at least when dry); crystals visible as short raised lines and dots on the younger leaves; primary lateral nerves 4–8 on either side; petioles mostly 1·5–4 cm. long. Racemes shortly pedunculate, ± 15–30(–40) cm. long, subdensely very many-flowered, usually crisped pubescent except often towards the base; bracts subulate to linear-lanceolate, 1–2 mm. long; pedicels 1·5–5(–10) mm. long, longest in long-staminate flowers, with small bracteoles near the middle or above. Flowers dimorphic, scented. Sepals triangular-lanceolate to ± oblong or ovate-oblong, bluntly pointed, 2–3 mm. long, slightly fleshy, pubescent at least along the margins, becoming reflexed. Long-staminate flowers white or cream; stamens (10–)12–15; filaments 3–4(–5) mm. long; anthers oblong-elliptic, 0·8–1·2 mm. long; carpels ± half as long as the filaments and only very rarely developing. Short-staminate flowers yellowish-green; stamens 8–15, with 1–1·5 mm. long filaments and rather smaller anthers; carpels (4–)5(–8), obliquely flattened-ovoid, united only at the base. Berry bluntly star-shaped in outline, the free part of the swollen carpels spreading, with the style-bases at the points, becoming more rounded, juicy and ripening orange or red, 5–8 mm. across. Seeds lenticular, 2·5–3(?–4) mm. across, papillate and striate towards the hilum, glossy black. Fig. 1.

UGANDA. Acholi District: Mt. Rom, Dec. 1935, *Eggeling* 2401!; Kigezi District: Kachwekano Farm, July 1949, *Purseglove* 3019!; Busoga District: Lolui I., 16 May 1964, *G. Jackson* U89!
KENYA. Turkana District: Murua Nysigar [Muruanisigar] Peak, 22 Sept. 1963, *Paulo* 992!; Aberdare Mts., Kabage, Apr. 1930, *Dale* in *F.D.* 2358!; Teita Hills, Ngangao Forest, 6 Feb. 1953, *Bally* 8760!
TANGANYIKA. Moshi District: Lyamungu, 8 Nov. 1943, *Wallace* 1125!; Kigoma District: Mugombazi [Mugombasi], 30 Aug. 1959, *R. M. Harley* 9457!; Songea District: Luwira-Kiteza Forest Reserve, 18 Oct. 1956, *Semsei* 2526!
DISTR. U1–4: K1–7: T1–8; widespread in tropical and South Africa, also in Madagascar

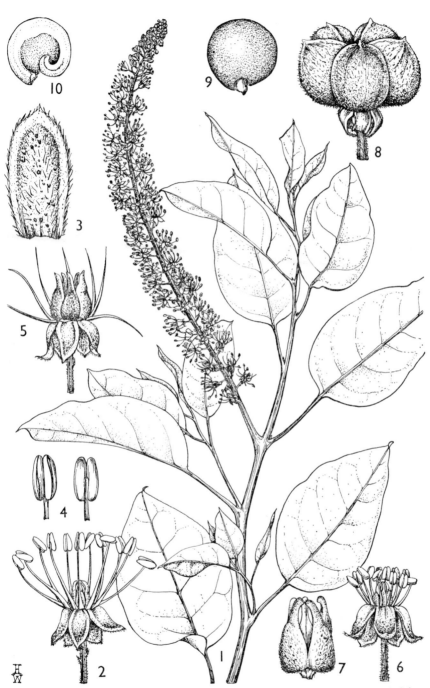

FIG. 1. *PHYTOLACCA DODECANDRA*—**1,** branch, with raceme of long-staminate flowers, × ⅔; **2,** long-staminate flower, × 4; **3,** sepal of same, outer surface, × 10; **4,** upper part of stamen, front and back views, × 10; **5,** exceptional long-staminate flower with developing carpels, × 4; **6,** short-staminate flower, × 4; **7,** gynoecium, × 10; **8,** fruit, × 4; **9,** seed, × 6; **10,** seed, with testa removed to show embryo and endo-sperm, × 6. 1–4, from *Tanner* 1781; 5, from *Polhill*; 6, 7, from *Bally* 8760; 8–10, from *Milne-Redhead* 1221.

HAB. Catholic, occurring in a wide range of forest, woodland, bushland, thicket and grassland communities, commonly riparian, generally perhaps on old forest land, limited only by extremes of altitude and rainfall; 500–2400 m.

SYN. *P. abyssinica* Hoffm. in Comm. Gotting. 12 : 25, t. 2 (1796); Oliv. in Trans. Linn. Soc. 29 : 140 (1875); Engl., Hochegebirgsfl. Trop. Afr.: 209 (1892) & P.O.A. C : 175 (1895), *nom. illegit.* Type: as for species
    P. abyssinica Hoffm. var. *apiculata* Engl., P.O.A. C : 175 (1895). Type: not specified
    P. dodecandra L'Hérit. var. *apiculata* (Engl.) Bak. & Wright in F.T.A. 6 (1) : 97 (1909); H. Walt. in E.P. IV. 83 : 44 (1909); V.E. 3 (1) : 137 (1915); F.D.O.-A. 2 : 255 (1938); T.T.C.L.: 450 (1949)
    P. dodecandra L'Hérit. var. *brevipedicellata* H. Walt. in E.P. IV. 83 : 44 (1909); F.D.O.-A. 2 : 255 (1938); T.T.C.L.: 450 (1949). Types: Tanganyika, Moshi, *Volkens* 1588 (K, isosyn. !) & Madagascar, *Hildebrandt* 3578 (K, isosyn. !)
    P. nutans H. Walt. in E.P. IV. 83 : 45 (1909). Type: Ethiopia, Simen, Ghaba valley, *Steudner* 557 (BM, K, iso. !)

NOTE. Notoriously poisonous, see Verdcourt & Trump, Common Poisonous Plants of East Africa.

The flowers are generally regarded as unisexual, but the reduction of the alternate organs is not very marked and the long-staminate or " male " flowers may sometimes develop fruits.

2. **P. octandra** *L.*, Sp. Pl., ed. 2 : 631 (1762); H. Walt. in E.P. IV. 83 : 58 (1909); Backer in Fl. Males., ser. 1, 4 : 231 (1951). Type: fig. 308 in Dillenius, Hortus Elthamensis (1732); no specimen OXF

A semi-succulent bushy herb 1–2 m. tall, glabrous on the vegetative parts. Leaf-blades elliptic-lanceolate or narrowly elliptic, (5–)7–18 cm. long, (1·5–)2·5–6 cm. wide, pointed and with a recurved mucro, cuneate, decurrent and unequal-sided at the base, very narrowly translucent at the margin; crystals visible as short raised lines and dots on the younger leaves; primary lateral nerves 6–10 on either side; petiole 1–3 cm. long. Racemes shortly pedunculate, 6–15(–23) cm. long, subdensely many-flowered, thinly crisped pubescent on the rhachis; bracts subulate or linear-caudate, 1–3 mm. long; pedicels 1–2 mm. long, with narrow bracteoles from the lower part. Flowers ♀, whitish or yellowish-green, the sepals sometimes flushed reddish in fruit. Sepals ovate-elliptic, rounded or bluntly pointed, 2·5–3(–3·5) mm. long, spreading, membranous, glabrous. Stamens 8–9, a little shorter than the sepals. Carpels (7–)8(–10), united, glabrous. Berry depressed-globose, only slightly lobed, with the style-bases around a median depression, 5–7 mm. across, ripening black or purplish-black. Seeds lenticular, 2–2·5 mm. across, smooth, glossy black.

KENYA. Uasin Gishu District: Eldoret, June 1958, *Gosnell* 687 !; Nairobi, Norfolk Hotel, 16 June 1956, *Kirrika* 249 !; N. Kavirondo District: W. Kakamega Forest Reserve, 13 July 1960, *Paulo* 551 !
DISTR. **K**3–6; native of tropical America from Mexico to Columbia, locally naturalized elsewhere in the tropics
HAB. Waste places; 1500–2100 m.

2. HILLERIA

Vel., Fl. Flum.: 47 (1825), Atlas 1, t. 122 (1835); H. Walt. in E.P. IV. 83 : 80 (1909)

Mohlana Mart., Nov. Gen. & Sp. 3 : 170, t. 290 (1832)

Herbs, often somewhat shrubby. Flowers in terminal and axillary racemes, ♀, zygomorphic, oblique. Sepals 4, 1 free, ± oblong-elliptic, the others united to ± the middle, with the rounded median lobe the longer, accrescent, surrounding the fruit, becoming prominently 3-nerved and some-times brightly coloured. Stamens 4(–13); anthers dorsifixed. Ovary of

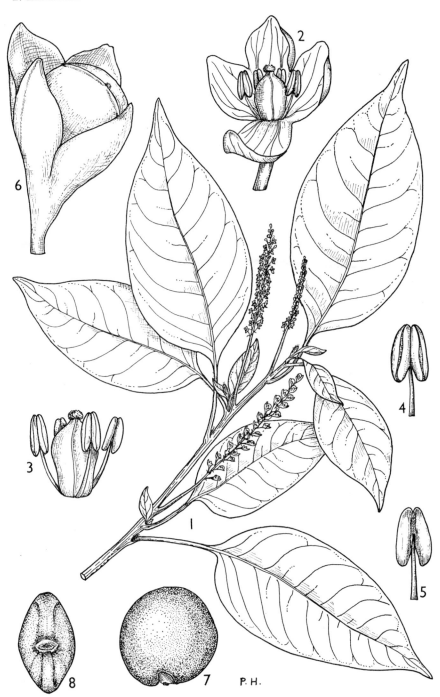

P. H.

FIG. 2. *HILLERIA LATIFOLIA*—**1,** fertile branch, × ⅔; **2,** flower, × 10; **3,** stamens and pistil, × 16; **4, 5,** upper part of stamen, front and back views, × 20; **6,** fruit, × 10; **7,** seed, side view, × 10; **8,** same, hilar view, × 10. 1, from *Drummond & Hemsley* 1820; 2–8, from *Dawkins* 419.

1 carpel, somewhat laterally compressed; style very short to as long as the ovary, tapered or capitate. Fruit discoidal, with a slightly toughened skin closely adhering to the seed, reticulately nerved when dry. Seed similarly shaped, with a brittle black testa.

Four species in South America, *H. latifolia* widespread also in Africa, Madagascar and the Mascarene Is.

H. latifolia (*Lam.*) *H. Walt.* in E.P. IV. 83: 81, fig. 25 (1909); V.E. 3(1): 137, fig. 88 (1915); F.P.N.A. 1: 143 (1948); T.T.C.L.: 449 (1949); F.P.S. 1: 107, fig. 68 (1950); Balle in F.C.B. 2: 98, t. 8 (1951); F.W.T.A., ed. 2, 1: 143 (1954); Cavaco in Fl. Madag., Phytolacc.: 4, fig. 1/1–3 (1954); F.P.U.: 106 (1962). Type: Mauritius, *J. Martin* (P, holo.)

Shrubby herb 0·4–2 m. tall, with some weak bristly pubescence along the angles of the youngest branches only. Leaf-blades ovate-elliptic to elliptic-oblong or elliptic-obovate, up to 8–16(–20) cm. long and 3·5–7·5(–9) cm. wide, markedly acuminate, broadly cuneate to the slightly decurrent unequal-sided base, glabrous except for small hairs on the nerves particularly beneath; crystals visible as short raised lines beneath; primary lateral nerves 6–10 on either side; petiole (1–)2–5(–7) cm. long, adaxially pubescent. Racemes ± 4–10(–25 in fruit) cm. long, subdensely many-flowered, setulose-pubescent along the ridges of the rhachis; bracts long-caudate from an ovate basal part, 1–2 mm. long, caducous; pedicels slender, 1·2–2(–4 in fruit) mm. long, with minute bracteoles on the upper part. Sepals green or white, often turning yellow or red in fruit, ± 1·5–2 mm. long in flower, accrescent to twice this size in fruit. Stamens 4, a little shorter than the sepals. Style very short or lacking, with an obliquely capitate stigma. Fruit 2–3 mm. across, laxly venose when dry, turning yellow to dark red or purple. Fig. 2.

UGANDA. Ankole District: Bunyaruguru, Feb. 1939, *Purseglove* 566!; Busoga District: Kityerera, Mar. 1931, *C. M. Harris* 14 in *F.H.* 81!; Mengo District: Kasa Forest, 4 Oct. 1949, *Dawkins* 419!
KENYA. Meru District: Thura Forest, June 1931, *Gardner* in *F.D.* 2608!; N. Kavirondo, Oct. 1931, *Jack* in *Napier* 145!; Masai District: Keekorok [Egalok]. 20 Oct. 1958, *Verdcourt & Fraser Darling* 2286!
TANGANYIKA. Bukoba District: Minziro Forest, *Willan* 224!; E. Usambara Mts., Kisiwani, 2 May 1939, *Greenway* 5880!; Morogoro District: about 8 km. NE. of Turiani, near Mtibwa, 26 Mar. 1953, *Drummond & Hemsley* 1820!
DISTR. U2–4; K4–6; T1, 3, 6; Ethiopia west to Liberia and Angola, Madagascar, Mascarene Is., South America
HAB. Rain-forest, riverine and groundwater forest; 450–1600 m.

SYN. *Rivina latifolia* Lam., Encycl. 6: 215 (1804)
    R. apetala Schumach. & Thonn., Beskr. Guin. Pl.: 84 (1827); P.O.A. C: 174 (1895). Type: Ghana, *Thonning* (C, iso.)
    Mohlana nemoralis Mart., Nov. Gen. & Sp. 3: 171, t. 290 (1832); Bak. & Wright in F.T.A. 6(1): 95 (1909); F.D.O.-A. 2: 252 (1938). Type: Brazil, State of Bahia, *Martius* (M, holo.)

NOTE. Generally regarded as an introduction to the Old World, but there seems no real reason for this supposition. It certainly seems native in some little disturbed East African forests, and if introduced must have become widely naturalized from an early date.

3. RIVINA

L., Sp. Pl.: 121 (1753) & Gen. Pl., ed. 5: 57 (1754), as *Rivinia*; H. Walt. in E.P. IV. 83: 101 (1909)

Herbs. Flowers in terminal and lateral racemes, ♀, regular. Sepals 4, free, oblong-elliptic or oblong-spathulate, somewhat accrescent, membranous, inconspicuously veined. Stamens 4; anthers dorsifixed. Ovary of 1 carpel,

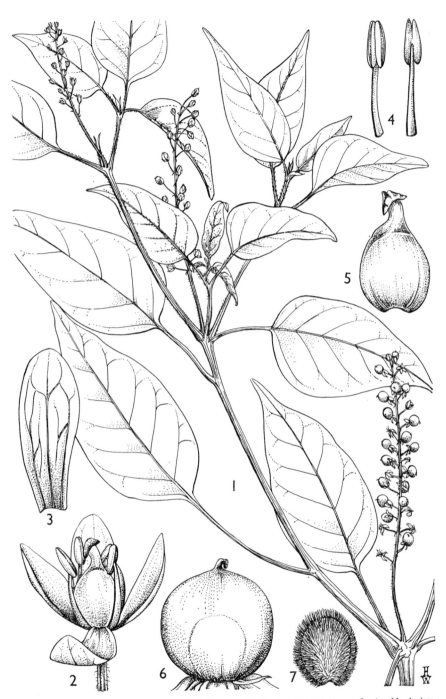

FIG. 3. *RIVINA HUMILIS*—**1**, habit, × ⅔; **2**, flower, × 15; **3**, sepal, × 20; **4**, stamen, front and back view, × 20; **5**, gynoecium, × 20; **6**, fruit, × 6; **7**, seed, × 6. 1, from *Fidalgo de Carvalho* 739 and a plant cultivated at Kew; 2–5, from former; 6, 7, from latter.

ellipsoid, somewhat flattened; style short, with a peltate stigma. Fruit a globose berry, with a thin pericarp, juicy. Seed discoidal, with a brittle black testa, often covered with very short spreading hairs.

An American genus of three species according to Walter in E.P. (1909), but Harms suggests, in a note under Heimerl's account in E. & P. Pf., ed. 2, 16C: 147 (1934), that these represent no more than one polymorphic species.

R. humilis *L.*, Sp. Pl.: 121 (1753); H. Walt. in E.P. IV. 83: 102, t. 30 (1909); Backer in Fl. Males., ser. 1, 4: 229, fig. 1 (1951); Cavaco in Fl. Madag., Phytolaccac.: 2, fig. 1/4, 5 (1954). Type: undetermined, not Hort. Cliff., nor Plumier, Nov. Pl. Am. Gen., t. 39 (1703)

Erect herb 3–10 dm. tall, glabrous to densely hairy on the branchlets, but commonly with short weak bristly hairs only on the ridges of the youngest parts. Leaf-blades lanceolate to elliptic-ovate or ovate, mostly 5–12(–17) cm. long, 2·5–6(–8) cm. wide, acuminate, broadly cuneate to rounded or cordate and unequal-sided at the base, glabrous to densely hairy, but commonly with small hairs only on the midrib above and on the nerves beneath; crystals often visible as short raised lines beneath; primary lateral nerves 6–12 on either side; petiole slender, 1·5–4·5(–6) cm. long, adaxially pubescent. Racemes 3–5(–11 in fruit) cm. long, sublaxly many-flowered; bracts caudate from a broader basal part, ± 1 mm. long, caducous; pedicels slender, becoming recurved, 1·5–3(–5 in fruit) mm. long, with minute bracteoles on the upper part. Sepals green, white or pinkish, ± 2 mm. long, spreading at anthesis. Stamens and ovary a little shorter than the sepals. Berry 3–4 mm. in diameter, ripening red. Seeds 2–2·4 mm. across, usually shortly hairy. Fig. 3.

Tanganyika. Lushoto District: Mombo, 9 Nov. 1955, *Milne-Redhead & Taylor* 7067! & 4 Aug. 1960, *Semsei* 3060!
Distr. **T3**; native of tropical and subtropical America, locally naturalized in the Old World tropics
Hab. In riverine forest and grassland; 550 m.

Note. Also grown as an ornamental, e.g. Kampala, *Stones* in *E.A.H.* H 5/56! and Nairobi City Park, *G. R. Williams* 620!; R. O. Williams reports it from gardens in Zanzibar (U.O.P.Z.: 434 (1949)).

INDEX TO PHYTOLACCACEAE